PASTURE MANAGEMENT FOR SMALL-SCALE FARMS

Optimizing Land, Grazing Systems, Livestock, And Sustainability For Efficient Farming

Jasper Mark S.I.

Table of Contents

INTRODUCTION 3

CHAPTER ONE 6
   SMALL-SCALE PASTURE MANAGEMENT 6

CHAPTER TWO 13
   PASTURE PLANNING AND DESIGN 13
   FORAGE SPECIES FOR SMALL FARMS 19

CHAPTER THREE 25
   GRAZING MANAGEMENT FOR SMALL LIVESTOCK 25
   INTEGRATED PEST AND WEED MANAGEMENT 30

CHAPTER FOUR 36
   WATERING SOLUTIONS FOR SMALL PASTURES 36
   SEASONAL CONSIDERATIONS 41

CHAPTER FIVE 48
   EQUIPMENT AND TOOLS FOR SMALL-SCALE PASTURE MANAGEMENT 48

CHAPTER SIX 55
   MARKETING PASTURE-BASED ITEMS 55

THE END 62

INTRODUCTION

I wanted to restore the beauty of the verdant fields in the peaceful countryside. I dreamed of returning these little farms to their former state of health and prosperity. I was going to accomplish this.

I started by doing some research. I attended classes, read books, and spoke with seasoned farmers. I gained knowledge about healthy soil, rotating animal grazing areas, and the significance of having a variety of grass species.

I then got to work tending to the land. I took my time tilling the ground, adding compost, and planting various grasses and plants. I was careful to sow each seed in its proper location. I enjoyed watching the fields go from being barren to being overflowing with life.

Maintaining these fields wasn't simple. I had to keep weeds out, check that the soil was sufficiently moist, and ensure that the animals were grazing properly. Although it was laborious, I gained insight into the requirements of the land.

Witnessing the return of animals to the area was one of the highlights. Butterflies circled the flowers, bunnies ran through the meadows, and birds built nests in the tall grass. This demonstrated to me that I was managing the land well.

People became aware of the wholesome fields I had created as the years went by. My secrets were sought after by other farmers. I was delighted to provide my knowledge to them.

I was proud of myself when I looked out at the fields after years of hard work. They were teeming with life, the soil was rich, and locals came to view and take in the stunning terrain.

My ultimate objective went beyond simply improving the fields for small farms. Restoring the land's health and leaving something positive for future generations were the main goals. As I stood there, gazing out at the verdant fields I had labored on, I realized that my dream had come true.

CHAPTER ONE

SMALL-SCALE PASTURE MANAGEMENT

Small-scale pasture management includes the care and maintenance of grazing land, commonly on a smaller scale, like a backyard or a small farm. While it may not be basically as immense as enormous-scale tasks, dealing with a pasture really is pivotal for the health and efficiency of livestock and the sustainability of the land. Here is a manual for small-scale pasture management.

1. Soil Health: Everything starts with the dirt. Test your dirt to comprehend its pH level, nutrient content, and piece. Change pH levels if it is important to

establish an ideal environment for grass development. Adding organic matter like compost can upgrade soil fertility and design.

2. Choosing Appropriate Grasses: Pick grass species that are appropriate to your climate, soil type, and expected use. Cool-season grasses like fescue and ryegrass are appropriate for calm climates, while warm-season grasses like Bermuda grass flourish in more sizzling locales. Consider the nutritional necessities of your livestock when choosing grass species.

3. Rotational Grazing: Rotational grazing includes splitting the pasture into smaller paddocks and rotating livestock between them. This forestalls

overgrazing, takes into account the rest and regrowth of grass, and amplifies forage usage. It likewise assists in making do with manuring dissemination, lessening parasite load, and further developing soil health.

4. Weed Control: Weeds can rival grass for nutrients and space, diminishing the quality and quantity of forage accessible to livestock. Standard cutting, manual evacuation, or designated herbicide application can assist with controlling weed populations without hurting helpful grass species.

5. Water Management: Satisfactory water access is fundamental for livestock health and pasture efficiency. Guarantee that your pasture has solid water

sources like lakes, streams, or boxes. Screen water quality to forestall tainting and guarantee that livestock approach clean drinking water consistently.

6. Fertilization: Ordinary fertilization can renew nutrients in the dirt and promote healthy grass development. Use composts in light of soil test suggestions to keep away from over application, which can prompt nutrient spillover and environmental contamination.

7. Pasture Redesign: Over the long haul, pastures might become corrupted because of compaction, erosion, or a decrease in grass power. Redesign strategies, for example, circulating air through, over seeding, and reseeding,

can restore pastures and further develop forage quality.

8. Bug Management: Bugs, pests, and sicknesses can harm grass and mischief livestock. Carry out coordinated pest management procedures that join biological, cultural, and chemical control strategies to limit bug harm while minimizing environmental effects.

9. Monitoring and Perception: Routinely screen your pasture for indications of nutrient deficiencies, weed pervasions, or different issues. Watch out for livestock conduct and health to measure the viability of your management practices.

10. Reasonable Practices: Aim for sustainable pasture management by minimizing chemical sources of information, rationing water assets, and promoting biodiversity. Executing environmentally agreeable practices helps your pasture and livestock as well as adds to the drawn-out health of the ecosystem.

Nonetheless, small-scale pasture management includes a mix of soil care, grass determination, rotational grazing, weed control, water management, fertilization, redesign, bug management, monitoring, and supportable practices. By executing these techniques, you can keep a healthy and useful pasture for your livestock while protecting the natural assets of people in the future.

CHAPTER TWO

PASTURE PLANNING AND DESIGN

Pasture planning and design are significant stages in guaranteeing the efficiency, health, and sustainability of a grazing framework. A very well-planned pasture gives more than adequate forage to livestock, promotes soil health, and limits environmental effects. Here are a few rules to consider while arranging and planning a pasture.

Site Assessment

Start with an exhaustive appraisal of your site. Think about the accompanying elements:

- Soil type and quality

- Geology and slant

- Climate and precipitation designs

- Existing vegetation

- Water accessibility and quality

Understanding these variables will assist you in drawing informed conclusions about pasture design and management practices.

Pasture Layout

While planning your pasture design, hold back nothing and be adaptable. Share the pasture into manageable paddocks or grazing units to work with rotational grazing. This considers the rest and recuperation of grazed regions, promoting healthier grass development and soil preservation.

Fencing

Great fencing is fundamental for powerful pasture management. Pick sturdy materials reasonable for your livestock and nearby circumstances. Electric fencing can be a cost-viable choice for partitioning paddocks and controlling grazing power. Guaranteed doors are decisively positioned for simple access and development of livestock and gear.

Watering Systems

Provide solid water sources inside every enclosure to guarantee livestock have ceaseless access to new water. This might incorporate boxes, lakes, or piped water systems. Think about the area, limit, and proficiency of your watering framework to address the issues of your herd and limit squander.

Forage Selection

Select forage species that are appropriate to your dirt type, climate, and livestock necessities. A different blend of grasses, legumes, and forbs can provide all-year forage, further develop soil fertility, and improve biodiversity. Counsel nearby agricultural augmentation administrations or forage experts for suggestions custom-fitting to your district.

Soil Fertility and Management

Keeping up with soil fertility is fundamental for supporting pasture efficiency. Direct customary soil tests to screen nutrient levels and pH, and apply proper composts or soil alterations depending on the situation. Carry out practices, for example, crop rotation, cover cropping, and organic matter addition, to further develop soil health and design.

Grazing Management

Implement an adaptable and versatile grazing management plan that thinks about occasional varieties, forage accessibility, and livestock nutritional necessities. Rotational grazing, strip grazing, and accumulating forage can assist with expanding pasture use, broadening the grazing season, and diminishing dependence on put-away feeds.

Monitoring and Adaptation

Routinely screen pasture conditions, livestock execution, and environmental effects to assess the viability of your management practices. Change your pasture plan and grazing methodologies depending on the situation in view of noticed results and evolving conditions.

In summary, pasture arrangement and configuration require careful consideration of site-explicit elements, livestock necessities, and management objectives. By implementing a thoroughly examined plan that focuses on soil health, forage quality, and effective grazing practices, you can create a useful and sustainable pasture system that benefits both your livestock and the environment.

FORAGE SPECIES FOR SMALL FARMS

For small farms, incorporating forage species can be a distinct advantage. These plants give significant nutrition to livestock as well as further develop soil health and lessen erosion. Picking the right forage species can have a tremendous effect on the efficiency and sustainability of a small farm. Here are some famous and helpful forage species that are reasonable for small farms:

1. Clover: Clover is a leguminous plant that is superb for further developing soil fertility. It fixes nitrogen from the environment into the dirt, diminishing the requirement for manufactured composts. Additionally, clover gives great forage to livestock, including cattle, sheep, and goats. It very well may be developed in a mix with grasses or as an unadulterated stand.

2. Ryegrass: Ryegrass is a cool-season grass that is exceptionally tasteful and nutritious for livestock. It tends to be grazed or gathered for feed. Ryegrass lays out rapidly and can endure an extensive variety of soil conditions. It is a phenomenal decision for winter

grazing or as a sidekick crop with legumes like clover.

3. Alfalfa: Horse feed is a lasting vegetable that is known for its profound root framework, which helps in separating compacted soil and further developing water penetration. It is a high-protein forage that is reasonable for dairy cows and other high-delivering livestock. In any case, horse feed requires all-around depleted soils and is really difficult to manage as compared with other forage species.

4. Orchardgrass: Orchardgrass is a cool-season grass that is known for its quick development and significant returns. It is acceptable for livestock and can be grazed or collected for roughage.

Orchardgrass performs well in an assortment of soil types and is lenient toward both the dry season and shade.

5. Timothy: Timothy is a cool-season grass that is appropriate for roughage creation. It has a high fiber content, making it reasonable for ponies and other livestock that require a high-fiber diet. Timothy is generally simple to lay out and manage, making pursuing it a well-known decision for small farms.

6. Bermudagrass: Bermudagrass is a warm-season grass that is known for its dry spell tolerance and high efficiency. It spreads by rhizomes and can rapidly cover uncovered soil, decreasing erosion. Bermudagrass is appropriate for grazing and roughage creation and

can likewise be utilized for erosion control in slanting regions.

7. Chicory: Chicory is a well-established, lasting spice that is dry, lenient, and exceptionally nutritious. It is rich in minerals and has been shown to further develop milk production in dairy cows. Chicory can be grazed or collected for feed and performs well in both cool and warm-season environments.

While choosing forage species for a small farm, it's fundamental to consider factors such as soil type, climate, expected use (grazing, roughage creation, and so on), and livestock inclinations. It's likewise helpful to pivot different forage species to further develop soil health, diminish irritation and infection pressure, and provide a nonstop inventory of excellent forage over time.

All in all, coordinating forage species into small farm activities can offer various advantages, including further developed soil fertility, erosion control, and excellent nutrition for livestock. By choosing the right forage species and carrying out legitimate management practices, small farmers can improve the sustainability and benefit of their activities.

CHAPTER THREE

GRAZING MANAGEMENT FOR SMALL LIVESTOCK

Grazing management for small livestock like sheep, goats, and poultry is essential for keeping up with the health of the animals and the efficiency of the pasture. Proper grazing practices can likewise help the environment by forestalling soil erosion and promoting plant variety. Here are a few essential standards and practices to consider while overseeing grazing for small livestock:

1. Rotational Grazing: Rotational grazing includes isolating the pasture into smaller paddocks and rotating the livestock between them. This permits

the vegetation in every enclosure to recuperate and regrow before the livestock return, forestalling overgrazing and promoting healthier pastures. Rotational grazing can likewise assist with controlling inner parasites by breaking the existence pattern of the parasites.

2. Stocking Density: The quantity of animals you have in a particular region, known as stocking density, ought to be carefully managed to forestall overgrazing. Overgrazing happens when animals consume more vegetation than the pasture can recover, prompting soil erosion and diminished pasture efficiency. Monitoring the state of the pasture and changing the stocking

density appropriately is fundamental for economical grazing management.

3. Rest Periods: Permitting the pasture to rest between grazing periods is fundamental for its recuperation and regrowth. Rest periods give the grass and different plants time to renew their energy and bounce back stronger. The length of the rest time frame will depend on variables like the sort of vegetation, weather patterns, and stocking density; however, a basic rule is to rest every enclosure for no less than 30 days between grazings.

4. Grazing Level: Monitoring the level of the grass and vegetation can assist with guaranteeing that the livestock are not grazing too forcefully. Leaving a

specific measure of leftover vegetation subsequent to grazing, commonly 3–4 inches, can assist with safeguarding the dirt from erosion and give ground cover to keep weeds from dominating. Changing the grazing level in view of the development rate and state of the pasture can assist with maintaining a healthy balance between grazing and regrowth.

5. Water and Shade: Giving satisfactory access to clean water and shade is fundamental for the health and prosperity of small livestock. Livestock ought to have simple access to water consistently, particularly in a sweltering climate when they need to remain hydrated. Shade can assist with

safeguarding the animals from outrageous temperatures and sun-related burns, diminishing pressure, and improving general efficiency.

6. Fencing and Infrastructure: Appropriate fencing is significant for overseeing grazing regions and controlling the development of livestock between paddocks. Electric fencing or transitory fencing can be utilized to create rotational grazing systems and separate larger pastures into smaller areas. Satisfactory infrastructure, like watering tanks, feeders, and shelter, ought to likewise be given to address the issues of the livestock and work with proficient management practices.

7. Monitoring and Observation: Normal monitoring and observation of the livestock and pasture conditions are fundamental for viable grazing management. This includes checking the body state of the animals, assessing the health of the pasture, and making changes in accordance with the grazing plan on a case-by-case basis. Keeping itemized records of grazing designs, stocking densities, and pasture conditions can assist in recognizing patterns and illuminate future management choices.

INTEGRATED PEST AND WEED MANAGEMENT

Integrated Pest and Weed Management (IPWM) is a comprehensive way to deal with overseeing pests and weeds on small-scale farms without depending exclusively on chemical pesticides and herbicides. It consolidates different techniques to keep up with crop health while minimizing environmental effects. Here is a basic manual for IPWM for small-scale farmers:

1. Crop Rotation: Turn crops in various regions of your farm each season. This helps break the existing patterns of pests and diseases, diminishing their development in the dirt.

2. Polyculture: Grow different crops together as opposed to a solitary crop. This befuddles pests and diminishes the possibilities of a vermin episode. It

additionally upgrades soil fertility and biodiversity.

3. Beneficial Bugs: Empower the presence of natural hunters and gainful bugs like ladybugs, lacewings, and ruthless parasites. They assist with controlling nuisance populations naturally. You can do this by establishing blossoms that draw in these bugs or by acquainting them with your farm.

4. Trap Crops: Plant trap crops that draw in pests from your primary crops. When the pests are attracted to these snare crops, you can undoubtedly manage and control them.

5. Physical Obstructions: Utilize actual hindrances like column covers, meshes, or fences to safeguard your crops from pests. This strategy is viable for forestalling pests like birds, bunnies, and a few bugs from getting to your crops.

6. Handpicking: Routinely review your crops for pests and weeds and eliminate them by hand whenever the situation allows. This is a serious technique, but it can be extremely compelling for small-scale farms.

7. Mulching: Utilize organic mulches like straw, leaves, or grass clippings to stifle weeds and hold dampness in the dirt. This additionally further develops

soil health and gives habitat to beneficial microorganisms.

8. Organic Sprays: Assuming that irritation or weed populations become excessively high, consider utilizing organic splashes produced using natural fixings like neem oil, garlic, or cleanser arrangements. These are less unsafe for the environment and can really manage pests and weeds.

9. Soil Health: Keep up with healthy soil by working on composting, cover cropping, and insignificant culturing. Healthy soil promotes strong plant development, making them more impervious to pests and infections.

10. Monitoring and Record-Keeping: Consistently screen your farm for indications of pests, infections, and weeds. Track pest and weed populations, as well as the adequacy of control techniques utilized. This will assist you with settling on informed choices and further developing your IPWM techniques over the long run.

CHAPTER FOUR

WATERING SOLUTIONS FOR SMALL PASTURES

Watering small pastures proficiently is fundamental for the health of your livestock and the efficiency of your territory. Here are some watering arrangements that are both pragmatic and cost-compelling.

1. Permanent Watering Tanks: Introducing super-durable watering tanks in essential areas inside your pasture can provide a solid water source for your livestock. These boxes can be associated with a primary water supply or filled manually, contingent upon your arrangement. Make a point to pick boxes

that are solid and simple to clean to keep up with water quality.

2. Portable Water Tanks: For smaller pastures or rotational grazing systems, versatile water tanks can be an adaptable arrangement. These tanks can be moved effectively to various regions of the pasture, depending on the situation. Consider tanks with float valves to guarantee a predictable water level and limit squander.

3. Pond or Stream Access: On the off chance that your pasture borders a lake, stream, or other natural water source, giving controlled access can be a basic watering arrangement. Introduce fencing or utilize impermanent electric fencing to direct livestock to the water

source while forestalling overgrazing or harm to the banks.

4. Gravity-Fed Systems: Gravity-fed watering systems utilize the natural power of gravity to circulate water to different areas inside your pasture. These systems can be set up utilizing stockpiling tanks raised over the ground, permitting water to stream downhill to boxes or waterers. This strategy can be particularly valuable in bumpy or slanted pastures where a customary water supply may be trying to be introduced.

5. Rainwater Reaping: Using water can be an eco-accommodating and cost-viable method for watering your pasture. Introduce drains and downspouts on

stables or different designs to gather water from tanks or barrels. This water can then be utilized to fill boxes or tanks, depending on the situation. Make sure to utilize a filtration framework to eliminate garbage and impurities before the water arrives at your livestock.

6. Automatic Waterers: Programmed waterers are a helpful choice that give a nonstop stockpile of new water to your livestock. These waterers are typically outfitted with float valves or sensors to maintain a reliable water level. While they might require a higher starting investment, programmed waterers can save time and work over the long haul.

7. Regular Maintenance: Whichever watering arrangement you pick,

standard maintenance is pivotal to guaranteeing appropriate capability and water quality. Clean boxes, tanks, and waterers routinely to forestall the development of green growth, microorganisms, and different impurities. Examine hoses, fittings, and valves for holes or harm, and fix or supplant any flawed parts quickly.

Note that there are a few down-to-earth and proficient watering arrangements accessible for small pastures. By choosing the right situation for your necessities and keeping up with it appropriately, you can guarantee a dependable inventory of clean water for your livestock while promoting the health and efficiency of your pasture land. Think about your particular necessities, financial plan, and the format of your pasture while picking a watering arrangement, and make it a point to seek advice from agricultural specialists or neighborhood extension services if necessary.

SEASONAL CONSIDERATIONS

Seasonal considerations play a significant role in pasture management for small-scale farms. Appropriate pasture management guarantees that livestock approach nutritious forage consistently, which thusly adds to their health, efficiency, and overall prosperity. Here are a few hints on overseeing pastures occasionally for small-scale farms.

Spring:

In spring, as the weather conditions heat up and grass starts to develop, it's fundamental to screen pasture development and address any dirt and nutrient deficiencies. Start by conducting soil tests to determine nutrient levels and pH. In view of the outcomes, apply proper manures to promote healthy grass development.

During the spring, rotational grazing can be helpful. Partition the pasture into smaller paddocks and pivot livestock routinely. This practice forestalls overgrazing, permits grass to recuperate, and promotes even forage development. It likewise helps in controlling parasites by breaking their lifecycle.

Summer:

Summer brings intensity and potential dry spell conditions, which can pressure pastures and diminish forage quality. To battle this, give satisfactory water sources to livestock and screen pasture conditions intently. Consider establishing dry-season-safe grass species and utilizing concealment designs to shield livestock from the intensity.

Proceed with rotational grazing all through the late spring months, yet be aware of grass level. Leaving grass too tall can prompt diminished quality, while grazing it too short can hurt the root framework. Expect to keep an ideal grazing level for healthy pasture development.

Fall:

As summer changes into fall, pasture development dials back, and grass starts to senesce. This is an ideal opportunity to get ready pastures for winter and guarantee sufficient forage accessibility. Lead another dirt test to survey nutrient levels and apply any fundamental changes.

Think about amassing grass by permitting it to develop in a specific region of the pasture without grazing. This gives forage to pre-winter and late-fall when new development is restricted. Proceed with rotational grazing;

however, diminish stocking rates to forestall overgrazing and soil compaction.

Winter:

Winter presents moves for pasture management because of cold temperatures, ice, and decreased forage accessibility. While it may not be possible to brush livestock on pasture throughout the cold weather for a very long time in colder locales, there are still ways of overseeing pastures successfully.

Limit access to delicate regions to forestall soil compaction and erosion. Give shelter and supplemental feed to livestock to meet their nutritional necessities. In the event that achievable, consider overseeding pastures with winter-solid grass species to further develop winter forage accessibility.

General Tips:

1. Monitor Pasture Health: Routinely investigate pastures for indications of overgrazing, weed invasions, and soil erosion. Address any issues quickly to keep up with pasture health and efficiency.

2. Water Management: Guarantee sufficient water sources are accessible all year. Appropriate hydration is fundamental for livestock health and forage development.

3. Fencing: Put resources into durable fencing to work with rotational grazing

and shield delicate regions from overgrazing.

4. Consult Nearby Assets: Contact neighborhood agricultural extension workplaces, colleges, or experienced farmers for counsel customized to your particular area and farm conditions.

Note that seasonal pasture management for small-scale farms requires careful preparation, monitoring, and flexibility. By executing these simple and functional tips, small-scale farmers can keep up with healthy and useful pastures all year, helping both their livestock and the land.

CHAPTER FIVE

EQUIPMENT AND TOOLS FOR SMALL-SCALE PASTURE MANAGEMENT

Small-scale pasture management requires a couple of fundamental pieces of equipment and tools to guarantee that the pasture stays healthy, useful, and liberated from weeds and pests. Appropriate management of pasture is critical for livestock health and the general efficiency of the land. Here are a portion of the fundamental hardware and devices you'll require for small-scale pasture management:

1. Mower or Farm Truck: A small work vehicle or a vigorous cutter is fundamental for cutting grass and weeds

in the pasture. Ordinary cutting keeps up with the level of the grass and keeps it from becoming congested, which can prompt low-quality forage and expanded weed development. Pick a cutter or farm vehicle that is suitable for the size of your pasture and simple for you to work on.

2. Drag Harrow or Chain Harrow: A drag harrow or chain harrow is utilized for separating manure heaps, fanning out manure equitably, and streamlining the outer layer of the pasture. This assists with further developing soil air circulation, water invasion, and nutrient appropriation. It likewise helps in controlling parasites by exposing them to daylight and air.

3. Fencing Materials: Great quality fencing is fundamental for overseeing livestock and controlling grazing regions. Depending on your necessities, you might require a wire network, electric fencing, wooden posts, and fencing instruments like pincers, wire cutters, and wall cots. Make a point to routinely check and keep up with walls to forestall escapes and keep hunters out.

4. Sprayer: A small sprayer or knapsack sprayer can be utilized for applying compost, herbicides, and pesticides to the pasture. It's critical to utilize these chemicals capably and as per the maker's guidelines to try not to hurt the environment or livestock.

Consider putting resources into organic or natural options whenever the situation allows.

5. Soil Test Pack: Customary soil testing is fundamental for understanding the nutrient levels and pH equilibrium of your pasture soil. A dirt test unit can assist you with deciding the right sort and measure of manure to apply, as well as distinguishing any dirt deficiencies that should be addressed. This guarantees that your pasture stays rich and useful.

6. Watering system: Admittance to perfect and fresh water is essential for livestock health and pasture efficiency. Depending on your arrangement, you might require a watering tank, a

programmed waterer, or a portable water tank. Try to routinely perfect and keep up with watering systems to forestall the spread of sickness and guarantee water quality.

7. Weed Control Tools: Hand tools like weed pullers, cultivators, and digging tools can be helpful for manually eliminating weeds from the pasture. For larger regions, you may likewise think about utilizing a weed trimmer or brush shaper. Incorporated weed management practices, including cutting, grazing management, and herbicide application, can assist with controlling weed development successfully.

8. Grazing Management Apparatuses: Devices like electrified barrier analyzers,

grazing sticks, and pasture rulers can help you execute and manage rotational grazing systems. Rotational grazing upgrades forage usage, further develops pasture health, and forestalls overgrazing.

9. Safety Gear: To wrap things up, consistently focus on wellbeing while working in the pasture. Put resources into individual defensive gear, for example, gloves, wellbeing glasses, and durable footwear. Make a point to follow all wellbeing rules and systems while working with hardware and dealing with chemicals.

Note that legitimate gear and apparatus are fundamental for successful small-scale pasture management. Putting resources into quality gear and rehearsing standard maintenance can assist you with keeping a healthy and useful pasture for your livestock. Continuously focus on economical and environmentally-accommodating management practices to guarantee the long-term achievement and resilience of your pasture ecosystem.

CHAPTER SIX

MARKETING PASTURE-BASED ITEMS

Marketing pasture-based items offers an exceptional chance for farmers and makers to take advantage of the developing shopper interest in reasonably delivered, top-notch food. Dissimilar to ordinary farming strategies that depend intensely on chemical data sources and restricted animal feeding tasks, pasture-based systems focus on animal government assistance, environmental stewardship, and nutrient-thick items. Here are a few techniques to successfully market pasture-based items.

1. Know Your Story

One of the most convincing parts of pasture-based items is the story behind them. Purchasers are progressively keen on knowing where their food comes from, how it's created, and the upsides of the individuals who produce it. Set aside some margin to articulate your farm's story—why you picked pasture-based practices, how you care for your animals, and what separates your items from others.

2. Highlight Benefits

Accentuate the various advantages of pasture-based items for expected clients. These can include:

• Higher nutritional worth: Pasture-raised meats and dairy items are in many cases higher in omega-3 unsaturated fats, vitamins, and cancer prevention agents compared with their traditionally raised counterparts.

• Animal government assistance: Feature the empathetic treatment of animals in pasture-based systems, accentuating the opportunity to wander and natural ways of behaving.

• Environmental advantages: Make sense of how pasture-based farming adds to soil health, biodiversity, and carbon sequestration.

3. Quality Over Quantity

Rather than concentrating exclusively on volume, underline the predominant quality of your pasture-based items.

Utilize graphic terms like "grass-fed," "pasture-raised," and "unfenced" to convey the quality and genuineness of your items. Consider getting affirmations from respectable associations to approve your cases and assemble entrust with purchasers.

4. Engage With Your People Community

Building strong associations with your nearby local area can be a strong marketing device. Participate in farmers' markets, have farm visits, or team up with nearby cafés and gourmet experts to grandstand your pasture-based items. Connecting straightforwardly with buyers permits you to teach them about the advantages of pasture-based farming and make steadfast clients who value your items.

5. Use Social Media and Web based Marketing

In the present computerized age, having a strong web-based presence is fundamental for contacting a more extensive crowd. Utilize web-based entertainment platforms like Facebook, Instagram, and Twitter to share your farm's story, post photographs and recordings of your items, and draw in clients. Consider making a site or blog to give more inside-and-out data about your farming practices, items, and forthcoming occasions.

6. Offer Sampling Opportunities

Permitting expected clients to test your pasture-based items can be a powerful method for prevailing upon them. Set up tasting stalls at farmers' markets, food celebrations, or neighborhood occasions to allow individuals an opportunity to try your meats, eggs, or dairy items. When they taste the distinction in quality and flavor, they're bound to become faithful clients.

7. Team Up With Similar Organizations

Partnering with different organizations that share your qualities can assist you in reaching new client segments and growing your market reach. Work together with nearby health food stores, centers, or eateries that focus on feasible, privately obtained items. Think about joint marketing endeavors, for example, facilitating cooking shows, making extraordinary menu items highlighting your pasture-based items, or offering packaged advancements.

Note that marketing pasture-based items requires a mix of storytelling, schooling, and local area commitment. By featuring the interesting advantages of your items, building strong associations with your neighborhood, utilizing advanced marketing channels, and working together with similar organizations, you can really promote your pasture-based items and draw in knowing customers who esteem quality, sustainability, and straightforwardness.

THE END

www.ingramcontent.com/pod-product-compliance
Lightning Source LLC
Chambersburg PA
CBHW030049230526
45471CB00003B/1007